Diabetes Management

Complete Guide and A Long-Lasting Solution to Diabetes Mellitus with the Help of Medication, Self-Monitoring and Healthy Diets.

Doctor Jane A. McCall

Table of Contents

Diabetes Management 1
INTRODUCTION 8
CHAPTER 1 9
   WHAT IS DIABETES? 9
CHAPTER 2 12
   WHAT IS TYPE 1 DIABETES 12
      Causes of Type 1 Diabetes 14
      Common Symptoms and Signs of Type 1 Diabetes. 16
      Life with Type 1 Diabetes 18
CHAPTER 3 20
   WHAT IS TYPE 2 DIABETES 20
      Causes of Type 2 Diabetes 22
      Common Diabetes Symptoms Associated With Type 2 Diabetes. 24
CHAPTER 4 27
   SELF-MONITORING OF BLOOD GLUCOSE 27
      What Is Blood Glucose Self-Monitoring? 28
      Who Should Self-Monitor Blood Glucose? 30

Target Blood Glucose Levels 32

How Is A Blood Glucose Monitor Used? 34

When Should Glucose Self-Monitoring Tests Be Done? 37

Real-Time Continuous Glucose Monitoring 39

CHAPTER 5 41

MANAGING DIABETES WITH DIET & FOOD PLANNING 41

What Diet Is Best For Diabetes? 42

Professional Help with Lifestyle Changes For Diabetes 47

Obesity, Diabetes and Diet 48

CHAPTER 6 50

MANAGING DIABETES WITH PHYSICAL ACTIVITY AND EXERCISE 50

Exercise and Diabetes 51

Types of Exercise For People With Diabetes 53

Monitoring Blood Glucose Levels When Exercising 60

Trying Out Blood Glucose Before, At Some Stage In, And After Exercising 61

Hypoglycemia and Exercise 63

When TO SEE A DOCTOR 65
Other Considerations 67
Recommendations 70
About the Author 74
Acknowledgments 76

Copyright © 2019 by Doctor Jane A. McCall

All rights reserved. No part of this publication may be reproduced, distributed, or transmitted in any form or by any means, including photocopying, recording, or other electronic or mechanical methods, without the prior written permission of the publisher, except in the case of brief quotations embodied in critical reviews and certain other non-commercial uses permitted by copyright law.

INTRODUCTION

Are you or someone you know suffering from diabetes or have recently been diagnosed with the condition? If you want to avoid the harmful and expensive pharmaceutical treatments that won't improve your health and discover natural methods that REALLY work then this is the book for you!

In this book you will be enlighten on the causes related to diabetes, symptoms related as for those undiagnosed to know what they are up against and the best possible way to manage diabetes with the help of healthy diet, Exercise and many more.

CHAPTER 1
WHAT IS DIABETES?

Diabetes, frequently known as diabetes mellitus, is a group of metabolic diseases in which the victim has excessive blood glucose (blood sugar), either due to the fact insulin production is inadequate, or due to the fact that the cells of the body's fail to respond well to insulin, or both. Patients with excessive (high) blood sugar will frequently experience polyuria (common urination), they may end up more and more thirsty (polydipsia) and increased hunger (polyphagia). If left untreated, diabetes can cause many complications. Acute complications can include diabetic ketoacidosis, hyperosmolar hyperglycemic state, or demise. Serious long term complications encompass cardiovascular sickness, stroke, chronic kidney disorder, foot ulcers, and harm to the eyes. Diabetes is due to either the pancreas not producing enough insulin or the cells of the body not responding properly to the insulin produced. The fundamental types of diabetes mellitus:

- Type 1 DM (diabetes mellitus): results from the pancreas's failure to produce enough insulin. This form was formally referred to as "insulin-dependent diabetes mellitus" (IDDM) or "juvenile diabetes".

- Type 2 DM (diabetes mellitus): starts off with insulin resistance, a situation in which cells fail to reply to insulin properly. as the disorder progresses a loss of insulin may additionally expand. This shape was formerly called "non-insulin-dependent diabetes mellitus" (NIDDM) or "person-onset diabetes".

- Gestational diabetes: is the third fundamental form and this do occurs when pregnant women without a preceding history of diabetes have high blood sugar levels due to one factor or the other.

CHAPTER 2
WHAT IS TYPE 1 DIABETES

Type 1 diabetes is an autoimmune ailment wherein the immune gadget destroys cells inside the pancreas. The body does no longer produce insulin which will help to fight against any form of infection or disease. Normally, the disease first appears in adolescence or early adulthood. Type 1 diabetes used to be called juvenile-onset diabetes or insulin-based diabetes mellitus (IDDM), however the disease could have an onset at any age. Type 1 diabetes makes up round 5% of all instances of diabetes.

In Type 1 diabetes, the pancreas is unable to produce any

insulin, the hormone that controls blood sugar levels. Insulin production turns into inadequate for the control of blood glucose levels due to the sluggish destruction of beta cells inside the pancreas. This destruction progresses without been aware over time until the mass of those cells decreases. To the extent that the quantity of insulin produced is insufficient.

Type 1 diabetes generally appears in childhood or youth, however its onset is likewise possible in maturity.

While it develops later in lifestyles, type 1 diabetes can be mistaken for type 2 diabetes. If correctly diagnose, it's miles referred to as latent autoimmune diabetes of maturity.

Causes of Type 1 Diabetes

The gradual destruction of beta cells in the pancreas that sooner or later consequences within the onset of type 1 diabetes is the result of autoimmune destruction. The immune machine turning against the body's own cells is probably brought on by means of an environmental component uncovered to human beings who've a genetic susceptibility.

Despite the fact that the mechanisms of Type 1 diabetes etiology are uncertain, they're thought to contain the interaction of more than one factor which are as follows:

- Susceptibility genes - some of which might be

carried by over 90% of patients with type 1 diabetes. A few populations - scandinavians and sardinians, for example - are much more likely to have susceptibility genes.

- Autoantigens - proteins thought to be launched or exposed at some point of ordinary pancreas beta cell turnover or Damage along with that caused by infection. The autoantigens set off an immune response resulting in beta cell Destruction.

- Viruses - coxsackievirus, rubella virus, cytomegalovirus, epstein-barr virus and retroviruses are amongst the ones which have been connected to type 1 diabetes.

- Eating regimen(Diet)- toddler exposure to dairy products, high nitrates in consuming water and low vitamin D intake have additionally been connected to the development of type 1 diabetes.

Common Symptoms and Signs of Type 1 Diabetes.

- Regularly feeling thirsty and having a dry mouth
- Changes on your urge for food, commonly feeling very hungry, occasionally even if you've currently eaten (this could also occur with weak point and problem Concentrating)

- Fatigue, feeling usually tired no matter slumbering and mood swings
- Blurred, worsening vision
- Gradual recuperation of skin wounds, common infections, dryness, cuts and bruises
- Unexplained weight changes, specifically dropping weight despite consuming the identical quantity (this occurs due to the body using alternative fuels stored in muscle and fat whilst releasing glucose in the urine)
- Heavy respiratory
- Potentially a loss of recognition
- Nerve damage that causes tingling sensations or ache and numbness in the limbs, feet and arms

(more common among people with Type 2 Diabetes)

Life with Type 1 Diabetes

Type 1 diabetes continually requires insulin treatment and an insulin pump or day by day injections could be a lifelong requirement to keep blood sugar levels under control. The situation was known as insulin structured diabetes.

After the prognosis of type 1 diabetes, health care providers should help patients discover ways to self-monitor via finger stick testing, the signs of hypoglycemia, hyperglycemia and different diabetic complication. Most patients will also be taught how to regulate their insulin doses.

As with other forms of diabetes, vitamins and physical hobby and workout are vital factors of the way of life control of the sickness.

CHAPTER 3
WHAT IS TYPE 2 DIABETES

Type 2 diabetes is the most popular form of diabetes, accounting for over 90% of all diabetes cases.

The number of adults identified with diabetes within the US has risen notably inside the past 30 years, almost quadrupling from 5.5 million cases in 2000 to more than 21.3 million in 2017.

Type 2 diabetes was once called adult-onset diabetes and Noninsulin-structured diabetes mellitus (NIDDM), however the ailment may have an onset at any age, increasingly more along with early life.

Type 2 diabetes mellitus most commonly develops in

adulthood and is more likely to occur in folks that are overweight and physically inactive.

Not like Type 1 diabetes which presently cannot be prevented, a number of the risk factors for type 2 diabetes may be modified. For lots of people, consequently, it is viable to save you the circumstance. Signs and symptoms that signal the need for diabetes testing:

- Frequent urination
- Weight reduction
- Loss of energy
- Excessive thirst.
- Slow healing of cuts
- Numbness or tingling in hands and feet

- Itchy skin

Causes of Type 2 Diabetes

Insulin resistance is normally the precursor to type 2 diabetes - a situation in which more insulin than normal is needed for glucose to enter cells. Insulin resistance inside the liver results in more glucose production at the same time as resistance in peripheral tissues means that glucose uptake is impaired. The impairment stimulates the pancreas to make extra insulin however sooner or later the pancreas is unable to make sufficient to save your blood sugar levels from growing too excessive.

Genetics performs a part in Type 2 diabetes - relatives of people with the disease are at a higher risk, and the Prevalence of the condition is much higher in particular amongst local Americans, Hispanic and Asian human beings.

Obesity and weight gain are vital elements that lead to insulin resistance and Type 2 diabetes, with genetics, diet, exercise and way of life all playing an element. Body fat has hormonal effects on the impact of insulin and glucose metabolism.

Once type 2 diabetes has been identified, health care

provider can help patients with a program of education and monitoring, including the way to spot the signs of hypoglycemia, hyperglycemia and other diabetic intricates.

As with other kinds of diabetes, nutrients, and bodily pastime and exercise are critical elements of the life-style Management of the situation.

Common Diabetes Symptoms Associated With Type 2 Diabetes.

Many people develop type 2 diabetes symptoms in midlife or in older age and gradually expand signs in stages, especially if the condition is going untreated and worsens. Type 2 diabetes signs and symptoms can consist of:

- Chronically dry and itchy skin
- Patches of dark, velvety skin inside the folds and creases of the body (normally in the armpits and neck). This is known as acanthosis nigricans.
- Common infections (urinary, vaginal, yeast and of the groin)
- Weight benefit, even without a change within the diet
- Ache, swelling, numbness or tingling of the hands and toes

- Sexual disorder, consisting of loss of libido, reproductive issues, vaginal Dryness and erectile dysfunction.

CHAPTER 4
SELF-MONITORING OF BLOOD GLUCOSE

Tight control of blood sugar levels is difficult to attain. Levels can fall too low despite the nice adherence to demanding day by day self-monitoring schedules.

The proportion of human beings inside the US with a diagnosis of diabetes who undertakes self-monitoring of glucose has risen dramatically - from 36% in 1994 to 64% in 2010.

All patients newly recognized with type 1 diabetes will obtain training on how to do their blood sampling and how to act on readings. Increasing numbers of peoples with type 2 diabetes - even those who do not need insulin treatment - also are encouraged to self-monitor their blood glucose levels.

What Is Blood Glucose Self-Monitoring?

The aim of self-monitoring is to accumulate exact facts about blood glucose levels through the years at multiple points. It helps to maintain regular glucose levels and prevent hypoglycemia, and permits the subsequent to be scheduled accordingly:

- The treatment regime/insulin doses
- Dietary intake
- Physical activity.

Such glycemic control is vital in the prevention of the long-term complications of diabetes.

In addition to monitoring diabetes treatment effects and identifying blood sugar highs and lows, self-monitoring is a method that guides overall treatment goals. Self-monitoring also gives perception into how diet, exercise and different factors, such as illness and stress, have an effect on blood sugar levels.

Self-monitoring allows patients enhance their knowledge of glucose levels and the effects of different behaviors on their blood glucose.

Patients on glucose-lowering drugs can take their self-monitoring information to their health care provider, allowing them to measure prescriptions accordingly and advise any modifications to diet and exercising.

Strict glycemic control in type 1 diabetes is tough to achieve - in spite of correct training on self-monitoring, the most common measurement does not provide enough information to avoid hypoglycemia.

Who Should Self-Monitor Blood Glucose?

It was formerly only people with insulin-treated diabetes- type 1 mainly - who would be recommended to self-monitor their blood glucose levels. Global guidelines now state that there is enough evidence for the benefit of glycemic control to suggest self-monitoring to anyone with diabetes, together with those with type 2 diabetes who do not need insulin treatment, as long as there is sufficient healthcare guide. Adequate support includes the following:

- The monitoring is included into an education program to promote appropriate treatment according to blood glucose values

- There is shared management with health care providers to offer a clear set of instructions for acting on results.

The type of diabetes determines how often self-measurement is needed. Type 1 diabetes demands several day by day measurements while insulin-treated type 2 diabetes needs only around two a day. If no insulin treatment is needed, much less than daily measurement can be sufficient.

Target Blood Glucose Levels

The general goal of glycemic control for adults with diabetes has been set by the American Diabetes Association, whose guidance is followed by health care providers. It states:

- The HbA1c level (a marker of average glucose levels over recent months) should be reduced to 7% to lessen the risk of diabetes complications

- If possible, and as long as hypoglycemia can be avoided, some individuals may be able to target an HbA1c of 6.5%.

Less ambitious HbA1c goals (which include getting below 8%) are appropriate for some patients, inclusive of the ones who have any of the following:

- History of severe hypoglycemia
- Limited life expectancy
- Advanced diabetes complications
- Extensive coexisting conditions.

Much less stringent targets may also be suitable for people with long-standing diabetes who discover targets difficult despite disease control strategies.

The 7% HbA1c level informs the equivalent self-monitoring targets that patients can aim for (and again, much less ambitious targets are appropriate for some patients):

- Before meals (preprandial) - 70-130mg/dl (3.9-7.2mmol/l)

- After meal (postprandial, 1-2 hours after start of meal) - less than 180 mg/dl

How Is A Blood Glucose Monitor Used?

A glucose meter electronically reads a small sample of blood on a test strip. The blood is commonly drawn by means of a pores and skin prick at the tip of a finger.

Over 20 varieties of glucose meter are commercially available, varying in size, the quantity of blood needed and electronic memory and analysis features. While a few permit graphs to be computed, for many it is up to the user to maintain meticulous information along with details of times, diet and exercise. Practical hints for blood glucose monitoring include:

- Handle the meter and test strips with clean, dry hands
- Use the test strips precise for the meter and keep these in the original container
- Use a test strip only once and discard
- Strips may be calibrated with the meter for accuracy, and some meters require coding with each new canister of strips
- Take a look at for expiration dates
- Keep in a cool, dry place

- Take the meter to office visits for checks by providers.

Practical steps are also needed in preparation of the skin prick for a blood sample. The skin site must be cleaned with warm, soapy water and dried, or an alcohol pad may be used. In any other case - if foods have been handled recently, as an instance - false readings can occur.

The lancet sizes vary and can be adjusted to prick the pores and skin one and produce the different amounts of blood needed by using various meters. Thinner and sharper lancets are typically the most comfortable. Lancets need to no longer be reused after single use.

To reduce pains, the sides of the finger may be used and fingers can be rotated, such as any of the 5 digits in preference to the index finger or thumb.

At the same time as the most accurate measurements are enabled via using the fingertips or outer palm, some meters permit using different sites consisting of the upper arms and thighs.

When Should Glucose Self-Monitoring Tests Be Done?

Individual instances of diabetes require different levels of blood glucose monitoring. The frequency of testing can change for an individual as well; the frequency may also need to be intensified within the event of changes to medications, stress levels, diet or activity levels.

Examples of the type of facts that can be provided via meter readings include checking oral medicines or long-acting insulin through the use of night-time fasting blood glucose (FBG) readings, taken at around 3 or 4am.

Test results from before ingesting can assist to guide changes to meals or medicines, and those obtained 1-2 hours following a meal are informative while learning how blood sugar levels are affected by foods. Tests at bedtime also help inform modifications to weight loss plan or medications.

Real-Time Continuous Glucose Monitoring

Non-stop glucose monitoring overcomes the hassle of taking several guide daytime readings from pores and skin pricks.

Patients with Type 1 diabetes usually do between four and eight finger-prick measurements every day, and rarely monitor night-time blood glucose levels.

Such self-monitoring can result in speedy modifications in blood glucose called excursion, together with postprandial hyperglycemia, asymptomatic hypoglycemia and fluctuations overnight.

Actual-time continuous glucose monitoring has been shown to be more powerful than self-blood glucose measurement in reducing hba1c in type 1 diabetes as it provide detailed information on glucose patterns and tends.

The most important crucial factor to the fulfillment of the devices is motivation and compliance of the consumer.

The available continuous monitors - some of which can be combined with insulin pumps-consist of an electrochemical sensor positioned below the skin and replaced every 3-7 days.

CHAPTER 5
MANAGING DIABETES WITH DIET & FOOD PLANNING

Alongside exercise, a wholesome diet program is a crucial detail of the life-style management of diabetes, in addition to being preventive against the onset of type 2 diabetes.

Maintaining a very good food regimen is likewise a crucial a part of maintaining tight control of blood sugar levels, itself important for minimizing the threat of diabetes complications.

The good news to people living with diabetes is that the circumstance does not preclude any specific type of food or require an uncommon food plan - the goal is much the same as it would be for everyone wishing to eat a healthy, balanced diet.

What Diet Is Best For Diabetes?

Having diabetes does not demand any particular difficult nutritional demands, and at the same time as sugary meals obviously have an effect on blood glucose levels, the weight loss plan does not have to be completely sugar-free.

Nutritional issues vary barely for peoples with different types of diabetes. For peoples with type 1 diabetes, weight loss plan (diet) is about managing fluctuations in blood glucose levels while for peoples with Type 2 diabetes, it is all about losing weight and restricting calorie consumption.

For people with Type 1 diabetes, the timing of food is particularly essential in terms of glycemic control and in relation to effect of insulin injection.

In general, however, a healthy, balanced diet is all that is needed, and the benefits are not limited to good diabetes control - they also imply good heart health. A healthy diet normally includes a variety of fruits and vegetable, whole grains, low-fats dairy products, skinless chicken and fish, nuts and legumes and non-tropical vegetable oils.

The following are some general dietary suggestions for a healthy life-style:

- Eat frequently - keep away from the effects on glucose levels of skipping food or having behind schedule meal due to work or long journeys (take healthy snacks with you)

- Eat vegetables and fruits and devour them in place of high-calorie meals - a variety of fresh, frozen and canned is good, but keep away from high-calorie sauces and meals containing added salt or sugar

- Whole grains high in fiber are recommended as a

healthy source of carbohydrate

- Consume pulses, a low-fat starchy source of protein and fiber, together with beans, lentils, chickpeas and lawn peas

- Reduce intake of saturated and trans fats by having poultry and fish without the skin and cooked, as an instance, under the grill, rather than fried

- Take a similar approach to cooking red meat while decreasing intake and looking for the leanest cuts

- Consume fish twice a week or more, however keep away from batters and frying - move for oily fish which include salmon, mackerel, sardine, trout and herring, which are rich sources of omega-3

- Keep away from partially hydrogenated vegetable oils and limit saturated fats and trans fats - replace them with monounsaturated and polyunsaturated fats

- Dairy awareness helps reduce fats intake - select skim (fats-free) milk and low-fats (1%) dairy products, reduce intake of cheese and butter and swap out creamy sauces for tomato-based ones

- Cut back on sugar by avoiding added sugars in drinks and ingredients - have tea and coffee without sugar, keep away from fruit that is canned in syrup and take note of food labels

- Cut back on salt - put together foods at home with

little or no salt and keep away from meals with high sodium such as processed foods

- Cut back on portion sizes - be cautious of quantities consumed when eating out

- Be cautious of "diabetic" foods - they are of no specific gain and can be expensive

- Drink alcohol only in moderation - as a guide, no more than one drink a day for women and no more than two for men.

Professional Help with Lifestyle Changes For Diabetes

In the US, the Community Preventive Services Task Force run diabetes prevention programs that help with enhancing diet for people at risk of, or newly diagnosed with type 2 diabetes. The program may consist of:

- Goals towards weight loss

- Individual and group education session on food regimen(diet) and exercise

- Meetings with diet and exercise counselors

- Individually designed diet and exercise plans.

Individuals in the national diabetes prevention program have access to a life-style coach to learn more about healthy eating and exercise.

Obesity, Diabetes and Diet

Obesity is a risk factor for Type 2 diabetes, and obesity

in people who already have diabetes results in bad control of blood sugar, blood pressure and cholesterol levels.

Some other situation with being overweight or having obesity is that it can worsen many of the complications of the diabetes.

Weight loss can be accomplished by following the tips above and limiting the consumption of calories.

CHAPTER 6
MANAGING DIABETES WITH PHYSICAL ACTIVITY AND EXERCISE

According to the Centers for disease control and prevention (CDC), over 29 million people in the United State have diabetes - a condition in which the body doesn't make enough insulin (type 1 diabetes), or is unable to use insulin properly (type 2 diabetes).

Insulin is a hormone, made in the pancreas, which regulates blood sugar (glucose) levels, and allows the body to use glucose for strength.

Exercise can help reduce complications of diabetes which

include:

- Heart sickness and stroke

- Blindness and other eye problems

- Kidney disease

- Amputations due to harm to blood vessels and nerves, leading to infection

A further 86 million people have pre-diabetes - a health condition that increases their risk of developing type 2 diabetes and different ailments.

Exercise and Diabetes

Aerobic sporting activities (exercises) such as brisk walking and hiking can also help to manage the onset of diabetes symptoms.

Preventing the onset of diabetes for people with pre-diabetes, or managing symptoms for those who have already got the condition, is crucial to keep health and prevent complications. Exercising is one proven way to help manage diabetes.

According to a joint position statement by the American College of Sports Medicine and The American Diabetes Association, exercise:

- Performs a key role in preventing and controlling blood sugar levels
- Can prevent or delay type 2 diabetes

- Can prevent diabetes in the course of being pregnant (gestational diabetes)

Staying physically active also helps prevent diabetes-associated health complications and improves overall quality of lifestyles.

Exercise is beneficial for people with diabetes as it improves insulin sensitivity by helping the cells of the body use insulin that is available. Physical activity additionally stimulates a separate mechanism, unrelated to insulin, to permit the cells to use glucose for energy, thereby regulating blood glucose levels.

Types of Exercise For People With Diabetes

The American Diabetes Association recommends two

types of physical activity for those with diabetes: ***Aerobic exercise and Strength training.***

Aerobic exercising

Also referred to as *cardiovascular exercise*, aerobic activity helps the body use insulin more efficaciously. It brings other benefits too, including:

- Strain remedy

- Progressed circulate

- Decreased danger of coronary heart disorder

- Lower blood strain

- Improved cholesterol levels

- Strong bones

- Weight management

- Better temper

Examples of cardio physical activities (aerobic exercise) consist of:

- Brisk strolling or trekking

- Low-effect aerobic exercise instructions

- Swimming

- Rowing

- Biking

- Basketball

- Dancing

- Skating

- Tennis

- Jogging

- Tai chi

How much aerobic activity is needed?

The president's council on health, sports and nutrition recommends:

Half-hour (30 minutes) daily of moderate physical aerobic activity as a minimum of 5 times weekly

This recommendation is for adults between the age 18-64. Adults with diabetes must have this in mind and work towards meeting this target.

People with a hectic schedule can also find it useful to do

numerous shorter exercises totaling 30 minutes daily - studies suggests that the benefits received are similar to the ones related to one longer exercise.

Strength training

Blood sugar levels can be reduced by strength training, such as using free weights.

Strength training, or resistance training, helps in lowering blood sugar levels and increase insulin sensitivity. In addition, it increases resting metabolism and builds stronger bones and muscle tissues, decreasing the risk of osteoporosis.

Examples of strength training consist of:

- Lifting free weights

- Lifting heavy objects, such as bottles of water or canned food

- Weight machines

- Resistance bands

- Exercise that use body weight such as sits-ups, squats, planks, and push-ups

- Strength training classes

How much strength training is needed?

Strength training ought to be undertaken at the least twice every week, furthermore to the recommended amount of aerobic activity.

Stretching sports

Stretching exercises are crucial for everyone, together with people with diabetes. Stretching:

- Reduces the chance of injury from aerobic exercise or strength training

- Increases flexibility

- Prevents muscle pain

- Lowers stress levels

Incidental physical activity

It can be useful to consider incidental bodily activity- everyday activity that are not classed as exercise however involve movement. A few studies show that such activities can make a contribution to improved fitness.

Types of incidental physical activities consist of:

- Taking the stairs instead of the elevator

- On foot (walking) to the bus stop

- Vacuuming

- Moderate intensity gardening

- Taking walks around the shopping center

- Washing the automobile

Monitoring Blood Glucose Levels When Exercising

To exercise adequately, many people with diabetes-especially those with type 1 diabetes or those on diabetes medicinal drugs - may need to check their blood glucose levels earlier than, at some point of, and after exercise.

This indicates how well the body is responding to exercise, and may help avoid blood sugar fluctuations, which can be dangerous.

Trying Out Blood Glucose Before, At Some Stage In, And After Exercising

Blood sugar levels must be tested half-hour before exercising. If they are:

- Lower than 100 milligrams per deciliter (mg/dl) - blood sugar may be too low to exercise. Low blood sugar is called hypoglycemia.

- Between 100 and 250 mg/dl - this is the most excellent range, within which it is far safe for the majority to begin exercising.

- 250 mg/dl or higher - blood sugar may be too high to exercise. Perform a urine test for ketones (which indicate more insulin is needed to control blood sugar). This is typically only a concern for those with type 1 diabetes.

In the course of exercise, in particular long exercises or new activities, blood sugar levels need to be examined every 30 minutes. Stop exercising if any of the subsequent symptoms are there:

- Blood sugar falls below 70 mg/dl

- Weakness

- Tingling

- Confusion

After exercise, check blood sugar level right away. Recheck levels numerous times over the following day – physical activity can lower blood glucose for as much as 24 hours.

Hypoglycemia and Exercise

If hypoglycemia (low blood sugar) is experienced during or after a exercising, it should be treated with straight away. This should be taking at least 15-20 grams of fast-acting carbohydrate such as:

- A sports drink

- Regular soda

- Glucose gel

- Jelly beans

Blood glucose levels should to be tested after 20 minutes, and the treatment repeated if they have not returned to normal. Follow the fast-acting carbohydrates with a protein such as peanut butter and crackers. Do not resume exercise until blood glucose returns to above 100mg/dl.

If hypoglycemia happens frequently throughout exercise, it can be important to adjust medications or the exercise regimen, or to absolutely consume a small snack before working out. Skipping food, strenuous exercise, or prolonged workouts can all cause hypoglycemia.

It should be noted that people with kind 1 diabetes are much more likely to experience hypoglycemia during or after exercise, even though people with type 2 diabetes may have problems if they are on medication for their condition.

When TO SEE A DOCTOR

For people with diabetes, it is far recommended to consult with a healthcare expert before any exercising programme commences.

It is miles advisable to consult a health practitioner

before beginning any new workout program.

A doctor can advise on the impact of medications on blood sugar levels all through the period of activities, and can provide a target range for blood glucose levels in the course of exercises. They will give the best advice to the perfect time to exercise, primarily based on the affected patients individual schedule, meal plan, and medication.

A doctor may additionally perform a physical check-up, looking at:

Heart health

Blood pressure

Diabetes-related complications

Depending on these complications, it may be

recommended to keep away from strenuous activities, or particular sports.

It is also important to consult a medical doctor if hypoglycemia is experienced often at some stage in or after exercise, or if any other undesirable side effects are experienced.

Other Considerations

Beginning an exercise plan can be daunting. It is far crucial to:

- **Set realistic goals** - start slowly - with just 5-10 minutes of exercise every day - and gradually increase the frequency and intensity of the activity.

- **Include aerobic and strength based activity** - an

exercise plan for diabetes management ought to include both aerobic exercise and strength training - studies shows that undertaking both forms of physical activity is more effective than doing just one of the two.

- **Take precautions** - constantly keep fast-acting carbohydrates on-hand in case of hypoglycemia. Consider wearing a clinical alert bracelet in case of emergency.

- **Pick foot-wears wisely** - many peoples with diabetes have issues with their feet, because of poor circulation and nerve damage. Wear comfortable and supportive jogging footwear.

- **Be consistent** - to reap the advantages of exercise for diabetes, it has to be undertaken regularly.

Recommendations

1) How and Where to Buy Viagra Online Safely, Legally and Cheap: The Secret Behind How To Buy Viagra Online Safely Without A Prescription (With List Of Best Place To Buy Viagra Online) http://getbook.at/viagraonline

2) Viagra & Sildenafil: Uses, Dosage, Side Effects and Risks Information: The Secret Guide Behind How To Buy Viagra Online Safely, Cheap and Legally (With Best Online Pharmacy for Generic Viagra) http://getBook.at/viagra

3) Erectile Dysfunction (ED): Symptoms & Causes, Diagnosis, Treatment Online, And More Using

Viagra Without A Prescription (Including Where To Buy Viagra, Cialis, Levitra etc Drugs Cheap & Safely Online

http://getBook.at/erectile

4) Innovative Visualisation: The Power of Mind Perception -- GET MORE DONE THROUGH MIND MANIPULATION, INCENTIVES, PSYCH TRICKS AND MORE

http://getBook.at/innovative

5) Natural Healing and Remedies Cyclopedia: Complete solution with herbal medicine, Essential oils natural remedies and natural cure to various illness. (The answer to prayer for healing)

http://getBook.at/naturalhealing

6) **100 BEST CAT WELLNESS FOOD, DIET & RECIPES**: The hidden healing power diet for cat kidney problems, cat weight-loss, & pregnant cat diet; including recipes for all cat diseases and illness http://myBook.to/catfood

7) **The Brain, Mind and Memory Therapy:** The Science of embracing Change, Boosting Brain Power, Increasing Your Energy and Mental Strength. http://getBook.at/brainbook

8) **What Wikipedia Can't Tell You About Achieving Your Goals**: Why your objective setting never works out the way you plan http://getBook.at/wikipedia

9) **The First Year From Childbirth and beyond:** Inside-out Information on what to expect the first year and beyond early childhood for mothers and fathers made simple

http://getBook.at/childbirth

We love Testimonies, and we want to know how thus our publications have been of immense help to you. And please consider writing to us at www.engolee.com

Follow us on Social media at:

Website: www.engolee.com

Facebook Page: www.facebook.com/engolee

Twitter Page: *www.twitter.com/engolee*

About the Author

Dr. Jane A. McCall is a willing Health Researcher who is committed to blessing human race. She has developed a series of fabulous and highly effective healthful strategies and exercise programs. She applies her encyclopaedic knowledge and astonishing perception to analyze the background and underlying causes of various diseases affecting people in the world and then designs individualized and totally effective strategies to attain the desired results in solving human related problem with diseases. Jane is totally committed to helping the world discover their ideal expression of complete wellbeing.

Acknowledgments

The Glory of this book success goes to God Almighty and my beautiful Family, Fans, Readers & well-wishers, Customers and Friends for their endless support and encouragements.

www.ingramcontent.com/pod-product-compliance
Lightning Source LLC
Chambersburg PA
CBHW021504210526
45463CB00002B/891